Spaceports

by Patrick H. Stakem

2017

1st edition

Number 14 in the Space series.

Table of Contents

Introduction..3
Author...3
Siting of Spaceports...4
U.S. Spaceports..4
    Kennedy Space Center...5
        Launch Complex 39...6
        Other KSC Launch Complexes...6
        The Mobile Launch Platform..7
        The Launch Control Center 7
        Shuttle primary and contingency landing sites...............8
    Edwards AFB, California...9
    Wallops Flight Facility 9
    Vandenberg AFB...10
    Mojave Spaceport...11
    White Sands Space Harbor..12
    Pacific Spaceport Complex, Alaska.......................................12
    Sounding rockets...12
Soviet Union/Russia..12
Japan..13
France..14
India...14
China...15
New Zealand..15
Brazil...16
Italy...17
Australia...17
Sealaunch...17
Sounding Rocket ports...18
Spaceports in Space..18
    Deep Space Gateway 19
    LOP-G..19
Wrap-up..19
Bibliography..21

Glossary of terms..24

Introduction

A port is a point of departure or arrival. This book covers the topic of Spaceports. Like an airport or a train station, these are hubs that people and cargo come to for a transportation system. Some launch sites allow for vertical launch, and some allow for horizontal take off and landing. These are indistinguishable from airports, although the runways are usually longer. Spaceports are the gateway to orbit for people and supplies. They are key to continued exploration of the solar system, and enhanced human exploration beyond the Earth. As we shall see, spaceports can be placed... in space. The Deep Space Gateway and the Lunar Orbital Platform - Gateway are examples of these. The Russians prefer the term Cosmodrome for the place from which rockets depart.

This book will not discuss purely military launch sites, and it is not intended to be exhaustive.

Author

Mr. Patrick H. Stakem has been fascinated by the space program since the Vanguard launches in 1957. He received a Bachelors degree in Electrical Engineering from Carnegie-Mellon University, and Masters Degrees in Physics and Computer Science from the Johns Hopkins University. At Carnegie, he worked with a group of undergraduate students to re-assemble, modify, and operate a surplus missile guidance computer, which was later donated to the Smithsonian. He was brought up in the mainframe era, and was taught to never trust a computer you could lift.

He began his career in Aerospace with Fairchild Industries on the ATS-6 (Applications Technology Satellite-6) program, a communication satellite that developed much of the technology for the TDRSS (Tracking and Data Relay Satellite System). He followed the ATS-6 Program through its operational phase, and worked on other projects at NASA's Goddard Space Flight Center including the Hubble Space Telescope, the International Ultraviolet Explorer (IUE), the Solar Maximum Mission (SMM), some of the Landsat missions, and Shuttle. He was posted to NASA's Jet Propulsion Laboratory for Mars-Jupiter-Saturn (MJS-77), which later became the *Voyager* mission, and is still operating and returning data from outside the solar system at this writing. He initiated and lead the international Flight Linux Project for NASA's Earth Sciences

Technology Office. He is the recipient of the Shuttle Program Manager's Commendation Award, and has completed 42 NASA Certification courses. He has two NASA Group Achievement Awards, and the Apollo-Soyuz Test Program Award. He has worked at both the Kennedy Space Center, the Wallops Flight Facility, and the Vandenburg launch facility.

Mr. Stakem has been affiliated with the Whiting School of Engineering of the Johns Hopkins University since 2007, and Capitol Technology University. Mr. Stakem supported the Summer Engineering Bootcamp Projects at Goddard Space Flight Center for 2 years

Siting of Spaceports

It's preferred to launch over water or sparsely populated land, for safety issues. Kennedy and Wallops sites launch east, over the Atlantic Ocean. Vandenburg launches south, to achieve polar orbit. If you launch east to a low Earth orbit, you take advantage of the Earth's rotation. If you launch from the equator, east, you take maximum advantage of this. The altitude of the site makes little difference, since most of the rocket's energy goes into velocity of the vehicle, as opposed to lifting.

Launch to geostationary orbits requires an orbital plane change, unless you are right on the equator. This can amount to 25% less mass to orbit. Since Sealaunch is mobile, it can launch efficiently to any orbital inclination.

U.S. Spaceports

This section discusses United States' facilities.

NASA uses several launch sites, and you can view them from a safe yet spectacular location. Before you go, don't assume the launch will take place at the appointed time. During pre-flight check-out, many problems are discovered that lead to delays. The launch will probably not occur at the posted time, or even week. Plan to be flexible.

Kennedy Space Center in Florida is still the go-to location for launches. Try to get there for one of the spectacular night launches. There is a designated viewing area.

Kennedy Space Center

The Kennedy Space Center was built adjacent to Patrick Air Force Base, which was a launch site for ballistic missile testing. Over 400 flights have left PAFB for Space. Kennedy has sent more than 165 rockets to orbit.

The Saturn moon rockets were assembled at the Cape vertically in the Vehicle Assembly Building at (VAB), on the Crawler-transporter, and checked out for flight. Before being erected on the crawler, each stage was inspected and tested after delivery. The stages had been checked out individually before shipment to the launch site. In the first three Saturn V flights, 40 serious defects were found and corrected at this point.

The VAB was built to integrate the Saturn-V rocket and the Apollo payload. It is the largest single story building in the world at over 500 feet tall, needing all that headroom to hold the erected Saturn moon rocket. It was completed in 1966, and was later used for assembly of the Space Shuttle missions.

Launch Complex 39 at KSC was used to launch the Lunar Missions. It has two launch pads, 39a and 39b, and connected with the Vertical Assembly Building via the crawler way. These were later used for the Space Shuttle Program.. The crawler was built by the Marion Power Shovel Company, of Marion, Ohio.

When the various stages of the Saturn-V vehicle were assembled in the VAB, the Apollo "stack" was connected to the Launch Control Center for checkout, via a high speed data line. After checkout, the crawler-transporter picked up the vehicle and its support base, and moved it to the launchpad. Here, the vehicle was connected to data lines leading from the pad to the LCC again.

The Launch Control Center at Kennedy Space Center had control of the Apollo launch until the vehicle cleared the launch tower, at which point control transferred to Mission Control in Houston. The LCC is Building 30 at KSC, located near the Vertical Assembly Building. It has two facilities called the Mission Operations Control Rooms. One is on the second floor, and the other is on the third floor. The lower floor was filled with mainframe computers and communications equipment. Room number 2, used for the Apollo Missions, including the Lunar Landings, was designated a National Historic Landmark and has been restored back to its original configuration.

Both in the VAB and at the pad, RCA computers in the LCC controlled the automated testing and verification. Tests are conducted from one of the two firing rooms in the LCC.

For checkout operations, a firing room would be occupied by up to 400 engineers, with additional support personnel in a backup room. There were 400 consoles, 100 of which had CRT displays. There were also four large overhead screens with projectors.

Launch Complex-5 was used for the early Redstone launches. Launch complex 14 was for the Atlas, and Launch complex 19 for the Titan-II. Launch complex 34 was used for the Saturn I and I-b, with LC39 dedicated to the Saturn-V, as well as STS (Shuttle).

Launch Complex 39

Launch Complex 39 at KSC was used to launch the Lunar Missions. It has two launch pads, 39A and 39B, and includes the Vertical Assembly Building and the connecting crawler way. These were later used for the Space Shuttle Program, and will be modified again for the next-generation Space Launch System, which is Shuttle-derived. Modifications to the facilities will be needed for the launch pad, the flame trench, the mobile launcher platform, and the Vertical Assembly Building. Recall that these facilities were built 50 years ago.

Other KSC Launch Complexes

Many of the launch facilities at Cape Canaveral were originally used by the Air Force for ballistic missile launches. Such vehicles as the Atlas and Titan, which were repurposed into satellite launch vehicles.

Launch Complex 36 was used for Atlas launches; but has been repurposed to launch Blue Origins' reusable boosters. There were two launch pads, A and B. They saw a total of 68 (A) and 77 (B) launches. The newer United Launch Alliance Atlas-V launches from Complex 41. This site was used for Titan-II and -IV vehicles previously. It will be used for the new, semi-reusable Vulcan vehicle.

Launch Complex 40 is located at the north end of the facility, and was used for Titan-III and -IV. It was leased to SpaceX in 2007 for the Falcon rockets, which have gone from the location 28 times. A Falcon-9 exploded at the facility in 2016 during a static test firing, and heavily damaged the facility. It was ready to operate again by the end of the year.

Complex-6, at the south end of the facility, was used for the early Redstone, Juno and Jupiter rockets. It shares a blockhouse with Complex-5. This facility was used for Allen Shepard's flight, the first American in space. Not to be forgotten, the chimpanzee Ham left for space from this facility. It saw a total of 23 launches.

The Mobile Launch Platform

The Mobile Launcher Platform, or crawler, carried the assembled Saturn-V rocket and Apollo payload from the Vertical Assembly Building 3 miles to the launch pad. The vehicle also included a Launch Umbilical Tower, a crane, and the water suppression system, which protected the assembly at launch. The platform on the crawler adjusted automatically to keep the vehicle vertical during the trip to the launch pad, which was on a 3% grade. It has been updated to support SLS, and is now called the Super Crawler. The crawlers had been modified from the Saturn to the Shuttle, and one crawler had been modified to support the Ares program, which is now canceled.

The general sequence of events is that a crawler is driven into the VAB, and the Solid Rocket Boosters are installed. Then, the liquid-fueled Core stage is added. After that checks out, the second stage is integrated. It is then ready for final checkout, and roll-out of the VAB. At the pad, there is a dry rehearsal, and a wet (fueled) rehearsal. After that, the vehicle is returned to the VAB to await launch.

The Launch Control Center

The LCC is Building 30 at KSC, located near the Vertical Assembly Building. It has two facilities called the Mission Operations Control Rooms. One is on the second floor, and the other is on the third floor. Room number 2, used for the Apollo Missions, including the Lunar

Landings, was designated a National Historic Landmark and has been restored back to its configuration at that time.

I was told that one must watch out for alligators in the area, mosquitoes frequent the area, and there is no shade. Minor matters, when there is a launch to see. Again, check the on-line schedule. Launches from KSC head to the East, out over the ocean. They can generally be seen from a large area up and down the coast from KSC. A large area to the east is cleared on the ocean and in the sky prior to scheduled launches.

Shuttle primary and contingency landing sites

The Shuttle landed without engines. It descended rapidly, and was said to have the aerodynamics of a brick. While it was intended to land at the runway at KSC, there were designated contingency landing sites. These stretched around the globe, to accommodate various launch and reentry contingencies. The purpose-built landing strip at KSC is 15,000 feet long. It was also used to accommodate the Shuttle carrier aircraft, a specially modified 747-freighter, that transported the Shuttle to the Cape when needed.

Edwards Air Force Base in California was also available, and featured not only concrete strips, but also the large dry lake beds surrounding the base. The first free flights of Enterprise landed at Edwards. White Sands, New Mexico was also a contingency site.

There were defined abort modes on the East Coast of the United States. None of these were ever used. They included Miami International; Myrtle Beach airport; Dover, Delaware; Halifax, Nova Scotia; St. Johns, Newfoundland; and a few others.

In the event of a Transoceanic abort, a series of airfields were certified for Shuttle landing. NASA personnel were stationed at these sites during a launch. These included Diego Garcia; Cologne, Germany; Casablanca, Morocco; the Gambia; Nigeria; Senegal in Africa; and Rota, Spain The list also included RAF base Fairford in the U.K.

A contingency site, if Kennedy and Edwards were both unavailable was White Sands Space Harbor in New Mexico. Some launches also occur from this facility.

The follow-on Space Launch System is a Shuttle-derived heavy lift vehicle with the job of getting large masses to orbit. It will require major modifications to Kennedy facilities.

Edwards AFB, California

Used for winged craft, Edwards AFB with its extensive dry lake beds was used as a contingency site for Shuttle landings. An X-15 rocket aircraft took off from there under the wing or a B-52. It was taken to altitude, then released, and it rode its engine beyond the defined limits of space. So, Edwards was a Spaceport. There were two spaceflights with the X-15.

Wallops Flight Facility

The Wallops Flight Facility (WFF) is located at Chincoteague, Virginia on the Atlantic Ocean. It supports sub-orbital and orbital flights. It is administratively under the Goddard Space Flight Center. It has a small museum and gift shop. It is possible to see a launch from Wallops – check their website for times. However, be aware, not everything goes smoothly at a launch, and the launch might be slipped hours, days, or be rescheduled. Besides orbital missions, Wallops launches sounding rockets from 6 launch pads, controlled from one of three blockhouses. There is also a large runway. Wallops can support launch operations from anywhere on the planet, thanks to its mobile range instrumentation, which can be delivered by aircraft.

Wallops has sent 19 payloads to orbit, and hundreds of sounding rockets to the edge of the atmosphere.

The Visitor's Center, Building J-20T, is located on Route 175, six miles east of US-13 and five miles west of Chincoteague Island, VA. It is free. https://www.nasa.gov/centers/wallops/visitorcenter

Also, get the NASA WFF ap. This keeps you up to date on what's happening, and when the next launch is scheduled.

Located to the south, the Mid-Atlantic Regional Spaceport is a commercial facility for launches. It is a joint project of the States of Maryland and Virginia.

Vandenberg AFB

Vandenburg Air Force Base in Lompoc, California, is the launch site for U. S. Polar orbiting missions. They launch south. The Air Force lets NASA and Space-X use the facility, but it is an active military installation, called the Western Test Range. It does not have a mosquito problem, but people have been killed off the beach by Great White Sharks.

It is also the site of the California SpacePort, a 100 acre commercial facility. Vandenberg has one of its launch pads on the National Registry of Historic Places. Vandenburg has launched over 700 flights for the Air Force, and NASA. It launched the first polar orbiting satellite, Discover-1, in 1959.

Vandenburg has facilities to launch Pegasus, Taurus, Minotaur, Atlas V, and Delta IV, as well as the Falcon-9, and ballistic missiles Atlas, Thor, Titan-I and -II, Minuteman, and Peacekeeper.

The Base was established in 1941, and currently hosts the USAF's 30th Space Wing, which includes the 30th Launch Group, the 30th Operations Group, and the 30th Mission Support Group. The facility also hosts the 9th Space Operations Squadron, the 21st Space Operations Squadron, the 148th Space Operations Squadron, a NASA Resident Office, and several other government units.

The Space and Missile Heritage Center is at Space Launch Complex 10, where the first IRBM tests of the Thor vehicle were made. Tours of this can be arranged by the 30th Space Wing Public Affairs Office. This Center contains displays from the Cold War.

The site was supposed to be used for the Air Force's Manned Orbiting Lab Project, which never flew. It was also to be the site for Shuttle launches to polar orbit, but these too were canceled. The launch pad for both of these projects was constructed. For supporting the Shuttle, the main runway was lengthened.

The Orbiter Maintenance and Checkout Facility was 22 miles from the runway, and a special vehicle was constructed to take it that distance. Boeing's X-37B a winged, reusable spacecraft, used the runway at Vandenberg.

At Kennedy, the Shuttle "stack" was assembled in the Vertical Assembly Building, and moved to the launch pad via the crawler. At Vandenberg, it was planned to integrate the Shuttle stack on the launch pad. A site for a contingency landing after a flight from Vandenburg AFB (which never occurred) was Easter Island.

Mojave Spaceport

The Mojave Air and Spaceport is located in California. It is licensed for horizontal launches of reusable spacecraft, and is certified as a spaceport. It functions as an airport, and includes three main areas of activity: flight testing, space industry development, and aircraft heavy maintenance and storage. It has hosted air racing events. Scaled Composites, the developer for SpaceShipOne and its carrier aircraft White Knight, is located on site, as are XCOR Aerospace, and Virgin Galactic. The first privately funded sub-orbital crewed flight took off from Mojave. The original Mojave Airport dates for 1935. It is also used for air racing contests, and has been the site of numerous aircraft world records.

Sometimes called the Civilian Aerospace Test Center, it is the first facility in the United States to be licensed by the FAA for horizontal launches of re-usable spacecraft. What this refers to is an air launch, where the first stage is a winged aircraft.

If you use an aircraft to take a rocket and payload to altitude, it is much cheaper than using a traditional first stage. This approach is used with Orbital Science's Pegasus vehicle. It can send nearly 1,000 pounds to low earth orbit. The "first stage" is a Lockheed L-1011 Stargazer. The approach is similar to that of the X-15 launch from a B-52. In theory, any airport with a sufficiently long runway could be used. In practice, only certain airports with low or non-existent passenger traffic are used.

Pegasus uses the air fields at NASA launch facilities, as well as the Kwajalein Missile Range in the Pacific Ocean, and the Canary Islands in the Atlantic.

White Sands Space Harbor

This facility is operated by NASA, as was used for training Shuttle pilots in approaches and landings. It was also a backup site for Shuttle landings, after the runway was lengthened. Mission STS-3, Shuttle Columbia, was the only one to make use of this feature. It is built on a dry gypsum lake bed, at a location called Alkali Flats.

White Sands was the launch site for the captured German V-2 missiles after World War-II. Seventy were available, and 47 of these were launched. It remains a launch site for sounding rockets.

Pacific Spaceport Complex, Alaska

The Pacific Spaceport Complex, in Kodiak Alaska, has also been used for launches to orbit. It opened in 1998, and has seen 17 launches as of this writing. It is located on Kodiak Island, and is operated by the Alaska Aerospace Development Corporation, for the State of Alaska. It is now termed the Pacific Spaceport Complex – Alaska. It has two launch pads.

Sounding rockets

The Poker Flats range in Alaska is the only high-latitude rocket range in the United States. It is also a tracking station. The Andoya Rocket Range is located on an island two degrees north of the Arctic Circle. It s owned and operated by Norway. The Esrange Space Center is located in northern Sweden It is 200 km north of the arctic circle, and handles sounding rocket and balloon payloads. It was built by ESA, and is owned by the Swedish Space Corporation. The Woomera Rocket Range in Australia is also the site of sounding rocket launches.

Soviet Union/Russia

The Baikonur Cosmodrome was the first Spaceport (1955), the launch site for the first satellite in orbit, and for the first human in space in 1961. Located in Kazakhstan, now independent, it is still used by Russia. Site

31/6 is used for manned Soyuz missions, as is site 1/5. Baikonur has seen over 140 crewed flights. In all, over a thousand payloads have left Baikonur for space. The LC-45 Zenit rocket is also handled by Baikonur, which is sited at 46 degrees North. As the Soyuz is currently the other crewed vehicles available, U.S. Astronauts join with Cosmonauts to go to and return from the International Space Station at Baikonur.

Another Russian spaceport is Kapustin Yar Cosmodrome, which has sent 85 payloads to orbit. It was established after World War-2, in 1946, and used for research with captured German V-2 rockets. It gained the title of Russia's Roswell, as it was the site of numerous UPO sightings. It has numerous launch facilities.

The Plesetsk Cosmodrome has had more than 1,500 launches to space. It is located some 800 km north of Moscow, and has been in use since 1957. It is a secondary facility, to Baikonur., but has taken on increasing responsibilities, since Baikonur is now located in a different country (Kazakhstan). It was originally used for ICBM tests by the military, especially for the R7 vehicle.

The former ICBM test site Svobodny was initially selected as a replacement for Baikonur Cosmodrome, but this was suspended in favor of a new facility, the Vostochny Cosmodrome, in the far east of the country, at the 51st parallel, north. Three rockets have been launched from there. It is about 500 miles from the Pacific Ocean. This center, when complete, will reduce reliance on Baikonur. Seven launch pads are planned for the facility for Soyuz, and a planned heavy lift vehicle.

Japan

The Tanegashima Space Center is a launch site to the south of the main islands. It is operated by JAXA, the Japanese space agency founded in 2003. The Space Center was opened in 1969. It handles integration and test, launch, and tracking. There are several additional smaller space centers. Tanegashima is ideally sited for launches to the south, for polar orbit. It has sent 65 payloads to orbit. It is used for the current H-IIA and H-IIB vehicles.

The Noshiro Testing Center has carried out engine development and testing since 1962. The Kakuda Space Center develops liquid rocket engines. The Tsukuba Space Center handles the space network. It is also the training center for Japanese Astronauts, and develops equipment for the Japanese module on the ISS. The Uchinoura Space Center is the launch site for the Epsilon rocket, which can put 590 kg into a sun-synchronous orbit.

A sun-synchronous orbit is a type of polar orbit, where the satellite passes over any given point on the planet's surface at the same time every 24 hour period. This is useful for spotting trends on the surface.

France

The Guiana Space Centre in Kourou, French Guiana, is 500 km north of the equator. The ESA launch facility there launches the Ariane rocket. Over 261 payloads have been sent to orbit from the site. It launches to the east. ESA pays two-thirds of the operating budget, and the center supports commercial launches as well. There are five launch pads, an assembly and test building, and the launch control center. ESA will also be able to launch its own Soyuz-2z vehicles from a new facility, some 27 km from Kourou. There is a detachment of the Paris Fire Brigand on site for fire safety. A unit of the French Foreign Legion provides base security. On the grounds is the old Devil's Island prison facility, which is now a tourist site, closed for geosynchronous launches.

France also used the Hammaguir Special Weapons Test Center in Algeria to launch Diamant-A rockets from 1957-1967. These have sent 4 payloads to orbit, and launched numerous sounding rockets.

India

The *Satish Dhawan Space Centre* in Andhra Pradesh, India, has sent 53 payloads to orbit. It also operates the Vikram Sarabhal Space Center. India's first venture into space was in 1963, when they launched a 2-stage U.S. Nike Apache. The first Indian-built rocket, the RH-75, flew in 1967. There is a Vehicle assembly building, with a second in construction. The base launches both sounding rockets, and orbit and interplanetary missions. The Spaceport is unusable from September to November due to monsoons.

One of the major contributions of the Indian Space Program has been the launch of 104 satellites on one rocket. These were limited-size Cubesats, but the launch established the concept of multiple Cubesats as a primary payload.

Besides a successful lunar orbiter, India sent a successful Lunar Mission to the moon, an a Mars orbiter to the Red Planet, the third nation to do so.

China

In 2003, the first Chinese crew to fly in space were launched from the Jiuquan Satellite launch center, located in the Gobi desert. This facility was built in 1958, as the the first of China's three spaceports. It is used for low and medium altitude orbits, with large inclination angles. It was built with Soviet assistance. The Chinese launched their first satellite in 1970, and their first crewed mission in 2003. There are 6 launch pads.

The Taiyuan Satellite Launch Center supports the Long March vehicle for launches of meteorological and Earth resources satellites into sun-sync orbits. There are three launch pads.

The Xichang Satellite Launch Center came online in 1984, and handles launches to geosynchronous orbits. It handles some launches for Intelsat. It also supports China's lunar exploration program.

The Wenchang Satellite launch Center has three launch pads. It was originally intended to support the manned launch program, but now uses the Long March rocket to send payloads to higher orbits.

Area 4 is the site for launches of the crewed Long March 2F. The rocket also uses the Taiyuan Satellite Launch Center, and the Xichang Satellite Launch Center.

New Zealand

The Rocket Lab Launch Complex is a commercial spaceport on the East side of the North Island. It is used to launch the company's Electron rocket with Cubesat payloads. It opened in 2016, and supported its first launch in 2017, which failed to reach its intended orbit. The next launch, in January of 2018, was successful.

RocketLab is an American Company focusing on Cubesats and SmallSats. They have a launch agreement in place with NASA. They launch from Great Mercury Island on the east coast of the North Island. They use their Electron launch vehicle. They are cleared for a launch every 72 hours for 30 years. They also have a suborbital sounding rocket, the Atea-1 (which is Maori for *space*). It has a 2 kilogram cargo capacity.

A Cubesat is a small, affordable satellite that can be developed and launched by college, high schools, and even individuals. The specifications were developed by Academia in 1999. The basic structure is a 10 centimeter cube, (volume of 1 liter) weighing less than 1.33 kilograms. This allows multiples of these standardized packages to be launched as secondary payloads on other missions. A Cubesat dispenser has been developed, the Poly-PicoSat Orbital Deployer, P-POD, that holds multiple Cubesats and dispenses them on orbit. They can also be deployed from the Space Station, via a custom airlock. The Cubesat origin lies with Prof. Twiggs of Stanford University and was proposed as a vehicle to support hands-on university-level space education and opportunities for low-cost space access. This was at a presentation at the University Space Systems Symposium in Hawaii in November of 1999.

Brazil

The Alcantara Launch Center is located on the country's North Atlantic coast. It is the closest launch site to the equator. It is owned by the Brazilian Space Agency, and operated by the Brazilian Air Force. The first launch, a sounding rocket, was in 1990. The Program is run by *Agência Espacial Brasileira* (the Brazilian Space Agency). Only sounding rockets are launched from this site, but plans are in place to accommodate the Russian Proton launch vehicle. There is a large runway, engine and payload preparation facilities, support for liquid fuel loading, a launch tower, a mobile integration tower and a control Center. They can launch the Brazilian VLS rocket with satellites to orbit.

Another launch site is at Barreira do Inferno ("Hell's Barrier"). It was constructed in 1965, and has seen some 225 launches. It can support launches form the Guiana Space Center, and Alcantara.

Italy

Italy uses the San Marco offshore platform at the Broglio Space Center off the east coast of Kenya. It has seen 9 launches to orbit. The facility was developed in a partnership with NASA. It accumulated 27 launches. It is no longer in use, but the site does have a tracking station.

Australia

Australia uses the Woomera Test Range, and has sent 2 payloads to orbit. It is operated by the Royal Australian Air Force. "Woomera" refers to a spear throwing device, in an indigenous people's dialect.

The Test Range is used for both military and civilian purposes. In the 1950's the Black Knight sounding rocket was launched form the facility. Australia's first satellite, WRESAT, was launched in 1967. the site has the distinction of having been designated as a National Engineering Landmark.

It has launched more than 250 Skylarks, a British-design sounding rocket, as well as 20 NASA Aerobees.

In a reverse maneuver, the Japanese satellite Hayabusa landed at Woomera, after visiting the asteroid 25143 Itokawa.

Sealaunch

Sealaunch is a unique commercial facility that allows equatorial launches from the Pacific Ocean. It has a ship, and a mobile marine launch platform called ocean Odyssey. It addresses commercial payloads, using the Russian Zenit-3SL rockets. It's track record was thirty launches, three failures, and a single partial failure by 2014. First launch was in 1999.

The ship, *Sea Launch Commander*, out of Long Beach, California, is the test and integration facility, and delivers the vehicle to the Ocean Odyssey launch platform. The launch platform is self-propelled. In theory, launches can be made from any location on any ocean's surface. In practice, safety concerns restrict the launch location. It's about 3,000 miles from Long Beach to the preferred launch location at 154 west longitude. During launch, the Command ship is 5 kilometers from the launch platform.

Sealaunch, llc, was a consortium of Norway, Russia, Ukraine, and the United States. The operation was going well up until 2014, with the Russian invasion of Ukraine. That is where the Zenit boosters are made, and the further availability of these was unclear. Since then, Sealaunch sold the company, and filed for bankruptcy. It emerged from bankruptcy, with help from a Russian company, Energia Overseas, Ltd. They are supposedly trying to find a buyer, due to the high $30 million dollar cost of infrastructure maintenance, even without any launches. S7 Group may buy the company.

It is possible to launch a satellite from a submarine, which give you a wide range of locations. The Russians did this in 1988, Launching a small Tubsat-N from the Barents Sea. There was a follow-on launch of the Kompass-2 satellite in 2006. The launch vehicle was the three-stage Shtil, using liquid propellant. This was a repurposed ballistic missile.

Sounding Rocket ports

A sounding rocket is a research vehicle launched on a sub-orbital flight. It is up and back, not going to orbit. Missions are generally in the range of 50-1,500 km, the lower levels also being the domain of balloon-borne payloads. Actually, a sounding rocket can be launched from a balloon, a combined device called a "blooster."

Sounding rockets are generally smaller than their orbital counterparts, and less complex to launch, thus, cheaper. They usually have a simple solid rocket motor. There are numerous sounding rocket launch sites around the world, and they can also be transported by helicopter to remote locations for a launch.

Spaceports in Space

You don't need to start from Earth. For example, the Apollo lunar landers launched up to lunar orbit to join the Command and Service modules, which returned to Earth. The International Space Station has the ability to (mechanically) launch Cubesats. They once re-purposed and instrumented an old Russian space suite, and kicked it out the airlock. It is referred to as "SuitSat."The space station does not support launching with chemical propulsion systems.

There is an cost advantage to launching to Low Earth Orbit, and using a dedicated Space Tug/upper stage to deliver a payload to Geosynchronous orbit. This was studied, but not implemented.

Deep Space Gateway

The Deep Space Gateway (DSG) was a NASA Project for a crewed station in cis-lunar space. It was intended as a jumping-off point. The Orion crewed vehicle is scheduled to be used for this effort. The Gateway would be located in a halo orbit around the Moon. By that, we mean that the spacecraft would be visible to Earth for its entire orbital path. The DSG would form an in-orbit ecosystem for missions to the lunar surface, and to Mars. Ion thrusters are proposed for station-keeping. These use electrical power for accelerating various (usually, inert) gasses to high velocity, rather than using fuel and oxidizer. The thrust is generally low, but can be sustained for long periods of time. The Gateway was going to be built for the Asteroid Redirect Mission. This mission was defunded in early 2017. It had the goal of rendezvous with an asteroid, achieving lessons-learned applicable to planetary defense. The DSG could serve as a jumping off point for lunar surface and Mars missions.

LOP-G

The Lunar Orbital Platform - Gateway is an update to the Deep Space Gateway, basically a name change and some technical details updated. It will serve as staging point, in lunar orbit, for the Deep Space Transport, a re-usable crewed vehicle for Mars missions, using electrical and chemical propulsion.

The facility will have a power and propulsion element (PPE) derived from the DSG, with a mass of 8-9 metric tons, and be capable of supplying 50 KW of solar electric power for the ion thrusters. The project is in early concept phase at this writing, but some modules are being designed. These include a cis-lunar habitation module, compatible with the Orion capsule, a gateway logistics module, with a robotics arm. The Gateway airlock module will allow EVA activities, and could also be used as a short-term habitat. Lessons-learned from the ISS will be used for these modules, and certain technologies and assemblies are being tested on the station.

Wrap-up

As more human activity is focused on lunar surface operations, there will be a launch site, or multiple sites on the moon. There is talk of putting a spaceport on Mars' moon Phobos, for refuel and resupply, without the penalty of having to launch from the deeper gravity well of Mars. Ok, naming rights for the first Jupiter spaceport, supporting exploration of the numerous moon, are up from grabs. And, at some time, we will need to launch a starship mission from somewhere...

Bibliography

Alexander, George *Moonport, USA,* 1977, ASIN-B00072LE6A.

Burleson, Daphne *Spacecraft Launch Sites Worldwide,* 2007, ISBN-0786424117.

Chiuli, Roy M. (ed) *International Launch Site Guide,* 1994, ISBN-1884989012.

Corliss, William R. *NASA Sounding Rockets, 1958-1968: A Historical Summary* (The NASA Historical Report Series), 2014, ISBN-1502793970.

Court, Darren White Sands Missile Range, 2009, ISBN-1531637868.

Deng, Zhou Feng Guang, *Gobi Sky Harbor: into the manned space launch site,* 2000, ISBN 780218441X.

DeVincent-Hayes, Nan; Bennett, Bo *Wallops Island* (Images of America), 2001, ASIN-B00945G006.

Eckles, Jim *Pocketful Of Rockets: History And Stories Behind White Sands Missile Range,* 2013, ISBN-1492773506.

Edwards AFB, *Edwards AFB- The and Now – A Pictorial Tour,* 2001, ASIN-B001BZOTXC.

Icon Group, *White Sands Missile Range: Webster's Timeline History,* 1916 - 2007, 2009, ISBN- 0546909272.

ISRO, *From Fishing Hamlet to Red Planet: India's Space Journey,* 2015, ASIN-B01965PJ32.

NASA, *America's Spaceport: John F. Kennedy Space Center,* 2014, ASIN-B00L82609O.

NASA, *Moonport, A History of Apollo Launch Facilities and Operations,* 1978, SP-4204, ASIN-B009FO02D2.

NASA Public Affairs, *The Kennedy Space Center Story*, Graphic House, revised Edition, 1991, ISBN-0961064854.

NASA Suborbital Projects and Operations Directorate: Sounding Rockets Program Office and Wallops Flight Facility Goddard Space Flight Center, *Rocket Program Handbook,* 1999, ASIN-B00CPJ8TLQ.

Newell, Homer Edward, *Sounding Rockets,* 1959, ASIN-B0007DQ6U8.

Pappalardo, Joe *Spaceport Earth: The Reinvention of Spaceflight*, 2017, Overlook Press, ISBN 1468312782.

Phillips, Lynne V. *KSC Vertical Launch Site Evaluation,* 2007, ASIN-B01FKLJOAA.

Popelínský, Lubomír and Bedrich Ruzicka, *Rakety A Kosmodromy,* 1986, ASIN-B00OF2WI0Y.

Seedhouse, Erik *Spaceports Around the World, A Global Growth Industry* (SpringerBriefs in Space Development), 2017, ISBN-

Stakem, Patrick H. *Visiting the NASA Centers*, 2017, ASIN-B0757ZVB2G.

Strom, S. *International Launch Site Guide, Second Edition*, 2006, ISBN-1884989160.

Umansky, Semyon P. *Launch Vehicles. Launch Sites*, 2003, ISBN-5941410085.

U. S. Army corps of Engineers, APOLLO LAUNCH COMPLEX 39 - FACILITIES HANDBOOK, 1968, ASIN-B002JSSQIO.

U. S. Government, *Histories of the Soviet / Russian Space Program - Volume 5: Soviet Space Programs: 1981-87 - Piloted Space Activities, Launch Vehicles, Launch Sites, and Tracking Support,* 2013, ASIN-B00CYUC1WE.

U. S. Government, *Encyclopedia of Military Space Operations at Cape*

Canaveral: From Early Ballistic Missile Launches in 1953 through Titan, Atlas, Delta, and EELV Launches, 2013, ASIN-B00DUPVR6I.

U. S. Government Accountability Office, *Space Shuttle: Issues Associated with the Vandenberg Launch Site*: Nsiad-87-32br, 2013, ISBN-1289292930.

Wallace, Harold D. *Wallops Station and the Creation of an American Space Program* (The NASA History Series), ISBN-1493625934.

West-Reynolds, David *Kennedy Space Center: Gateway to Space*, 2010, Firefly Books, ISBN- 1554076439.

World Spaceflight News, *1961 to the 21st Century: Launch Complex 39 Then and Now - KSC Apollo and Shuttle Facilities,* 2001, ISBN-1893472388.

Glossary of terms

AEB – Brazilian Space Agency (Agência Espacial Brasileira)
AFB – Air Force Base.
AOA – Abort once around – Shuttle launch abort mode.
ASIN – Amazon Standard Inventory Number
Blooster – balloon launch of a rocket.
CATC – Civilian Aerospace Test Center (Mojave)
CCAFS – Cape Canaveral Air Force Station.
CLA - Centro de Lançamento de Alcântara, Alcantara Launch Center (Brazil)
CNES – (French) National Centre for Space Studies
Cosmodrome – Russian word for Spaceport.
CRT – an early display device using a cathode ray tube.
Delta-V – change in velocity
DoD – (U. S.) Department of Defense
ECAL – East Coast abort landing (Shuttle)
ESOC – European Space Operations Centre
ESRO – European Space Research Organization
ESTRACK – European Space Tracking Network.
FAA – (U.S.) Federal Aviation Administration
GTO – geostationary transfer orbit.
HTHL – horizontal take-off, horizontal landing
HTVL - horizontal take-off, vertical landing
IRBM – Intermediate Range Ballistic Missile.
ISBN – International Standard Book Number.
ISRO – Indian Space Research Organization.
JAXA - Japan Aerospace Exploration Agency
JSLC – (China) Jiuquan Satellite Launch Center
KSC – Kennedy Space Center, Florida.
LC – launch complex, pad, blockhouse, and workshops
LC-17 A & B – at KSC, for Thor and Delta.
LC-34 & 37 – at KSC, for Saturn-I and In
LC-39 – at KSC, Saturn-V and Shuttle.
LCC – KSC Launch Control Center
MST – mobile service tower.
NASA – (U. S.) National Aeronautics and Space Administration
NASDA – National Space Development Agency of Japan
Nippon Delta – Delta launch vehicle, built under license by Mitsubishi.

O&C – ops and control
OCST – (U.S.) Office of Commercial Space Transportation, part of the
FAA – (U.S.) Federal Aviation Administration.
PAFB – Patrick Air Force Base (at KSC)
PSLV – (India) Polar Satellite Launch Vehicle
SLF – Shuttle Landing Facility
SLS – Space Launch System.
STS – Space Transportation System (Shuttle).
TAL – Transatlantic abort scenario for the Shuttle
TNSC – (Japan) Tanegashima Space Center
USAF – United States Air Force
VAB – Vertical Assembly Building, KSC.
VAFB – Vandenberg Air Force Base
WFF – Wallops Flight Facility, Virginia
WSSH – White Sands Space Harbor

If you enjoyed this book, you might also enjoy one of my other books in the Space series. Most are also available in printed edition as well.

Stakem, Patrick H. *The History of Spacecraft Computers from the V-2 to the Space Station*, 2013, PRRB Publishing, ISBN-1520216181.

Stakem, Patrick H. *The Saturn Rocket and the Pegasus Missions, 1965*, 2013, PRRB Publishing, ASIN-B00BVA79ZW.

Stakem, Patrick H. *Spacecraft Control Center*, 2015, PRRB Publishing, ASIN-B01D1Y5LZ0.

Stakem, Patrick H. *Embedded Computer Systems for Space*, 2015, PRRB Publishing, ASIN-B018BAYCCM.

Stakem, Patrick H. *Microprocessors in Space*, 2011, PRRB Publishing, ASIN-B0057PFJQI.

Stakem, Patrick H. *Apollo's Computers*, 2014, PRRB Publishing, ASIN-B00LDT217.

Stakem, Patrick H. Crewed *Spacecraft*, 2017, PRRB Publishing, ISBN-1549992406.

Stakem, Patrick H. *Crewed Space Stations*, 2017, PRRB Publishing, ISBN-1549992228.

Stakem, Patrick H. *Rocketplanes to Space*, 2017, PRRB Publishing, ISBN-1549992589.

Stakem, Patrick H. *Visiting the NASA Centers*, 2017, PRRB Publishing, ISBN-154965120X.

Stakem, Patrick H. *Deep Space Gateways*, 2017, PRRB Publishing, ASIN-B077XJZQ1Y .

Stakem, Patrick H. *Manufacturing in Space*, 2017, PRRB Publishing, ASIN-B078KJ2RVQ,

Stakem, Patrick H. *NASA's Ships and Planes*, 2017, PRRB Publishing, ASIN-B078SH4P82.

Stakem, Patrick H. *Space Tourism,* 2018, PRRB Publishing, ASIN-B078TLTPVB.

Stakem, Patrick H. *Rocket Science-101*, PRRB Publishing, ASIN-B079474PD6.

Stakem, Patrick H. *In-Space Robotic Repair*, 2018, PRRB Publishing, ISBN-9781977066701.

Patrick H. Stakem, *Lunar Orbital Platform-Gateway*, 2018, PRRB Publishing, ISBN-1980498628.

Patrick H. Stakem, *STEM-Sat, Using Cubesats in the pre-K to 12 Curricula, A Resource Guide for Educators*, 2017, ISBN-1549656376.

Space suits, 2018.

Cloud Robotics, 2018.

Launch Vehicles, 2018.

Orbital Debris, 2018.

www.ingramcontent.com/pod-product-compliance
Lightning Source LLC
Chambersburg PA
CBHW032311240526
45464CB00023BA/2988